Sponsored Technology Workshops 2021

Held at the 2021 AIChE Annual Meeting

Boston, Massachusetts, USA and Online
7 - 11 November and 15 - 19 November 2021

ISBN: 978-1-7138-5721-1

Printed from e-media with permission by:

Curran Associates, Inc.
57 Morehouse Lane
Red Hook, NY 12571

Some format issues inherent in the e-media version may also appear in this print version.

Printed with permission by Curran Associates, Inc. (2022)

For permission requests, please contact AIChE
at the address below.

AIChE
120 Wall Street, FL 23
New York, NY 10005-4020

Phone: (800) 242-4363
Fax: (203) 775-5177

www.aiche.org

Additional copies of this publication are available from:

Curran Associates, Inc.
57 Morehouse Lane
Red Hook, NY 12571 USA
Phone: 845-758-0400
Fax: 845-758-2633
Email: curran@proceedings.com
Web: www.proceedings.com

TABLE OF CONTENTS

638309 Understanding the challenges facing Chem-E Education & Assessing Career Readiness of Students

Abstract:

In addition, we will seek feedback from attendees to understand current challenges and share possible solutions.

637764 Application of CFD-DEM in the Process and Pharmaceutical Industries

Abstract:

In this workshop, you would learn about Rocky DEM and the expanding range of granular systems that can be simulated accurately and effectively. Combined with CFD (Computational Fluid Mechanics), many processes across multiple industries can be simulated while representing real particle shape, particle size distribution and an accurate account of particle dynamic.
During the workshop presentation, industrial example and case studies will be used to highlight applications areas including:
• Slurry mills
• Cyclones, desanders, and drill cutting removal
• Pneumatic conveyors, hydraulic conveying
• Grain drying and sorting
• Tablet and candy coating
• Biomass reactors
• Fluidized beds and catalytic reactors

638841 State of Art in Modeling and Modeling Deployment for the Process and Pharma Industries; from Advanced Simulations to Web Apps

Abstract:

Taking the drug manufacturing process as an example, the presentation will highlight the breadth and value of M&S when it comes to manufacturing. Starting from upstream processes like mixing and clean-in-place, to downstream processes like chromatography, separation, and fill/finish, and ending with packaging and shipping, the value of M&S will be highlighted. Furthermore, the concept of simulation democratization will be introduced where advanced simulations will be made accessible to plant/manufacturing engineers using simple to use Web-Apps."

637762 Novel Altair Simulation Solutions for Particulate and Particle Laden Flows

Abstract:

Altair EDEM is a discrete element method solver which enables users to accurately simulate particulate processes. Coupled to Altair Computational Fluid Dynamics (AcuSolve) this also enables the simulation and analysis of two-phase solid-fluid systems.

These simulation tools are widely used in various industries and academia to deepen the understanding of different unit operations such as mixers, coaters, fluidized beds and pneumatic conveyors. EDEM and coupled AcuSolve-EDEM provide particle-level information about processes that can be impossible or extremely challenging to obtain through experimental studies. Despite the great potential of these tools for investigating various processes, there are some shortcomings when employing them.

In this workshop, a brief overview of EDEM and coupled AcuSolve-EDEM is presented. A few industrial test cases performed by EDEM and AcuSolve-EDEM are demonstrated. Some shortcomings associated with these tools and DEM and CFD-DEM simulation techniques in general are discussed. These challenges include long computational time, the size and number of particles which can be simulated in a reasonable computational time, particle shape and the material calibration process. The unique and novel solutions that Altair offers in order to address these challenges when using DEM and coupled CFD-DEM in various industries are presented.

637980 Implementing and Accelerating Autonomous Closed-Loop Control

Abstract:

Bayesian Optimization is a new machine learning method that can accelerate the path from automation to autonomous manufacturing. We will discuss the benefits of Bayesian Optimization, how it compares to more traditional methods such as APC and MPC, and how it can be used to optimize a real fermentation process. More broadly, we will show how this new technology enables accelerated DoE, product development, and fully autonomous manufacturing with closed-loop AI control.

637769 Role of physics based models for digital twins

Abstract:

Models are imaginary representations of the world that we build to understand it. They help us navigate the complexity involved in the naturally occurring physico-chemical processes. This can be utilized to design engineering solutions that are necessary for as a civilization such as power generation, chemical manufacture and others.

Models can be created based on recorded data (using ML and data analytics) as well as starting from fundamental principles of physics. Building physics based models that capture details and then translate them in to reduce models (rather than other way around) that can be utilized during operations is a prudent way to ensure safety and avoid errors.
The advancement and consistent adoption of cutting-edge digital simulation is critical to this future, and necessary in order to respond to the concerns of the market and of the public. Predictive modeling and simulation (M&S) techniques for reactor thermal hydraulics (TH) are essential to bringing the desired improvements to the economics and safety standards of process & power plants.

In this workshop, starting from the high level concept of digital twins for chemical and nuclear operations, we will dive in to the details of design and eventually capturing flow and turbulence behavior. This will set the importance of accurate physics based models for creating such digital twins. IT will be explained with the help of examples. We will also talk about the tools necessary to accomplish this as well as how they need to be integrated in to a common engineering process.

638245 Bridging the Knowledge Gap by Empowering Engineers with IIoT Advanced Analytics

Abstract:

You've been gathering production process data for years, but are you leveraging its full potential? Enabling your operational experts to turn data into actionable information is crucial in accelerating growth, controlling product quality and increasing sustainability.

Whether your goals are related to energy management, waste reduction, or increasing uptime, having your operational data at your fingertips is critical.

In this presentation, we will demonstrate how the practical use of self-service data analytics allows process engineers and operators to leverage all of their historical and time-series data - without having to rely on data scientists and spreadsheets.

AUTHOR INDEX

Notes:

Notes:

Notes: